DISCARD

Radio

by Darlene R. Stille

Content Adviser: Professor Sherry L. Field, Department of Social Science Education, College of Education, The University of Georgia

Reading Adviser: Dr. Linda D. Labbo, Department of Reading Education, College of Education, The University of Georgia

Compass Point Books

Minneapolis, Minnesota

Compass Point Books
3722 West 50th Street, #115
Minneapolis, MN 55410

Visit Compass Point Books on the Internet at *www.compasspointbooks.com* or e-mail your request to *custserv@compasspointbooks.com*

Photographs ©: Gregg Andersen, cover; Pictor/Charles Gupton, 4; Sally Vanderlaan/Visuals Unlimited, 6; David Clobes, 8; Photo Network/M. Bednar, 10; Marilyn Moseley LaMantia, 12; Photo Network, 14; NBC Photo/Archive Photos, 16; Photo Network/Dennis MacDonald, 18; Photo Network/T. J. Florian, 20.

Editors: E. Russell Primm and Emily J. Dolbear
Photo Researcher: Svetlana Zhurkina
Photo Selector: Phyllis Rosenberg
Designer: Melissa Voda

Library of Congress Cataloging-in-Publication Data
Stille, Darlene R.
    Radio / by Darlene R. Stille.
       p. cm. — (Let's see)
    Includes bibliographical references and index.
    1. Radio—Juvenile literature. [1. Radio.] I. Title.
TK6550.7 .S75 2001
621.384—dc21   2001001448

© 2002 by Compass Point Books
All rights reserved. No part of this book may be reproduced without written permission from the publisher. The publisher takes no responsibility for the use of any of the materials or methods described in this book, nor for the products thereof.
Printed in the United States of America.

Table of Contents

Listen to the Radio 5

How Your Radio Works 7

Let's Visit a Radio Station 9

Broadcasting a Program 11

Music in the Air 13

The First Radio Broadcast 15

Old-Time Radio 17

Two-Way Radio 19

Other Uses for Radio 21

Glossary 22

Did You Know? 22

Want to Know More? 23

Index 24

Listen to the Radio

Do you have a radio at your house? People everywhere listen to radios. We have stereo radios that play through big speakers. We have clock radios that wake us up in the morning.

We have small portable radios we can take to the beach. And we can listen to some radios through earphones.

Most cars also have radios. You can listen to the car radio while you ride around.

◄ *Portable radios are very popular.*

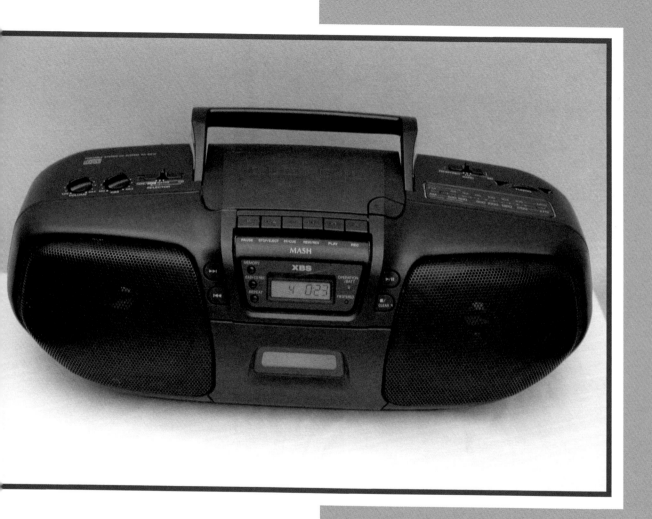

How Your Radio Works

All radios need electricity. Portable radios need batteries. You plug other radios into electrical outlets.

A switch or a button lets you turn your radio on. The volume control dial or button lets you play your radio loudly or softly.

A dial or button lets you tune your radio to your favorite radio station. Every radio station has a number on the dial.

◄ *Buttons and dials help us operate radios.*

Let's Visit a Radio Station

Radio stations **broadcast**, or send out, radio programs. Let's meet some of the people you hear on the radio.

A disk jockey plays music and talks about the songs. An announcer reads the news and the weather reports. A talk-show host interviews people.

The people you hear on the radio speak into **microphones**. A microphone changes sounds into electrical signals.

◄ *A radio announcer talks into a microphone.*

Broadcasting a Program

Electrical signals from a microphone travel through wires to a tall tower. The tower is called an **antenna**.

There, the electrical signals get changed into radio waves. Radio waves travel through the air. The antenna sends radio waves out in all directions.

Radio stations can broadcast AM or FM radio waves. AM radio waves travel farther than FM radio waves. FM radio waves are clearer, though. Some radios let you listen to either AM or FM stations.

◀ *Tall towers send signals out through the air to radios.*

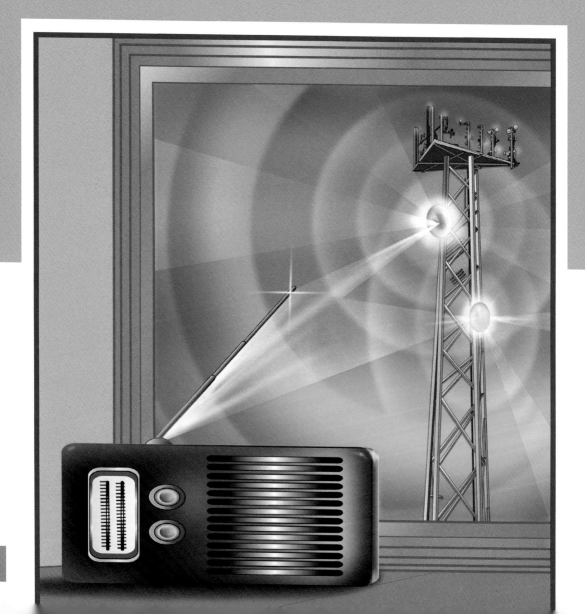

Music in the Air

The air all around you is filled with radio waves. Radio waves are outdoors. Radio waves are indoors.

You cannot see radio waves. You cannot feel radio waves.

But you can hear sound from radio waves when you turn on your radio. Your radio has an antenna that "picks up" radio waves.

◄ *The antenna picks up radio waves.*

The First Radio Broadcast

An Italian inventor sent the first sounds on radio waves in 1895. His name was Guglielmo Marconi. The sounds were clicks called **telegraph code**. The code carried messages.

Telegraph messages sent by radio were very important for ships at sea. People on sinking ships called for help. Other ships heard the radio signal and came to rescue them.

◀ An early 20th century telegraph operator sends a message using Morse code.

Old-Time Radio

Soon, inventors learned how to make radio waves carry music and voices. Families started to buy radios. The first radios were very big. Some of them sat on the floor.

In the 1920s, radio stations began broadcasting dance bands, soap operas, and other shows. Families gathered around the radio to enjoy their favorite programs.

◄ Actors broadcasting a live radio show during the 1940s

Two-Way Radio

Radio broadcasts are one-way radio. You can listen but you can't talk. Two-way radio lets two or more people talk to each other.

Police and fire fighters use two-way radio. Airplane pilots use two-way radio to help them take off and land safely.

Taxi drivers get two-way radio messages telling them to pick up passengers. Truck drivers use radios to talk with one another on the road.

◄ *Two-way radios help fire fighters do their jobs.*

Other Uses for Radio

Radar is a system that uses radio waves to "see" airplanes and other things that are far away or hidden in fog. The images on a radar screen come from radio waves that bounce off the objects. Radio waves can cook things too. Microwave ovens use powerful radio waves.

Radio waves carry phone messages across oceans and sometimes to and from spacecraft. Radio telescopes on Earth "listen" for radio waves coming from deep space.

Radio waves help people communicate. With radio, we can stay in touch.

◀ Radio telescopes explore objects millions of miles away in outer space.

Glossary

antenna—a tower that sends and receives radio or TV signals
broadcast—to send out
microphone—an instrument that changes sound into electronic signals
radar—a system that finds objects in deep water or fog
telegraph code—a system for sending messages over long distances
volume—the loudness of sound

Did You Know?

- Many amateur radio operators use radio transmitters and receivers to talk to one another. These people are often called "ham radio operators" or "hams."
- Radio waves travel through air at 186,000 miles (300,000 kilometers) per second!

Want to Know More?

At the Library
Birch, Beverley, and Robin Bell Corfield (illustrator). *Marconi's Battle for Radio*. New York: Barron's Juveniles, 1996.
Oxlade, Chris. *Radio*. Chicago: Heinemann Library, 2001.
Parker, Steve. *Guglielmo Marconi and Radio*. New York: Chelsea House, 1995.

On the Web
A Science Odyssey: Radio Transmission Activity
http://www.pbs.org/wgbh/aso/tryit/radio/
Includes an activity showing how radio works

Inventors Museum
http://inventorsmuseum.com/radio.htm
For information on how radio was invented

Through the Mail
National Public Radio
635 Massachusetts Avenue, N.W.
Washington, DC 20001
To learn more about this radio network and to find your local public radio station

On the Road
The Museum of Radio and Television
465 North Beverly Drive
Beverly Hills, CA 90210
or
25 West 52nd Street
New York, NY 10019
To learn about old-time radio broadcasts and visit a working radio studio

Index

AM radio waves, 11
announcers, 9
antennas, 11, 13
broadcasts, 9, 19
car radios, 5
clock radios, 5
controls, 7
disk jockeys, 9
earphones, 5
electrical signals, 9, 11
FM radio waves, 11
Marconi, Guglielmo, 15
microphones, 9
portable radios, 5, 7
radar waves, 21
radio stations, 7, 17
radio telescopes, 21
radio waves, 11, 13, 17, 21
speakers, 5
stereo radios, 5
talk-show hosts, 9
telegraph code, 15
two-way radios, 19

About the Author
Darlene R. Stille is a science editor and writer. She has lived in Chicago, Illinois, all her life. When she was in high school, she fell in love with science. While attending the University of Illinois, she discovered that she also enjoyed writing. Today she feels fortunate to have a career that allows her to pursue both her interests. Darlene R. Stille has written more than thirty books for young people.